PENNSYLVANIA VOICES BOOK VII

Living Literacy through Technology and Music to Develop Self-Efficacy in Computer Enhanced College English Composition Classes

Maryann Pasda DiEdwardo

authorHOUSE®

AuthorHouse™
1663 Liberty Drive
Bloomington, IN 47403
www.authorhouse.com
Phone: 1-800-839-8640

First published by AuthorHouse   9/18/2009

ISBN: 978-1-4389-6221-4 (sc)

Library of Congress Control Number: 2009903042

Printed in the United States of America
Bloomington, Indiana

This book is printed on acid-free paper.

Dedication

To my students who have generously allowed me to use their comments and writing to teach and to write; I applaud your talents!

"Most importantly, music as a catalyst helps the student find a way to relate to the literary message and encode language through sound."

Maryann P. DiEdwardo

Contents

I. Preface

Statistical results of my case study research have suggested that pairing music and linguistic intelligences in the college classroom improves students' grades and abilities to compose theses statements for research papers in courses that emphasize reading and writing skills (DiEdwardo 2004). Grounded in educational theories, as well as offering practical applications, pairing two intelligences advances student potential. Based on multiple intelligence theory (Gardner 1993), the Mozart Effect (Campbell 2001), and participation as precedent to learning (Bateson 1975), the conclusion may be drawn that integrating music into elementary, middle school, high school, advanced placement, and college linguistic classrooms, as well as into virtual classrooms enables students to learn to read and write (DiEdwardo 2004).

Student Comments

Spring 2003 English 151C-15

Student#1   I thought music made it easier to connect with the literature. I came to class excited and ready to do assignments and projects. It is easier to connect modern music that we know to literature that we aren't too aware of. It was really easy to do papers and I enjoyed that. I learned to take my time. Reading poetry, fiction, and drama improved because I understand better through music.

Spring 2003 English 151C-27

Student #1   Music is so inspirational and such a different way of communication. This makes for thought provoking writing. The first or main sentence in a song is just like the thesis. Music is a good example of how to construct a piece of writing and makes me want to come and sit through class more when I know that I can select my own music. Music helps me understand where the writer of literature finds material.

Student #2 Music helped bring out deep thoughts. Music is healing and can touch every mood the rhythm can move. I enjoyed this class and wanted to come. Music helped me to be more creative. I used cut and paste method to create assignments. Music helped me go into my thoughts process quicker. Every type of writing in English is easier with the help of music.

Summer 2003 English 101C-22

Student #1 I really enjoyed having music around. Even in the background when I'm writing, music helps me think better. My sentence structure became better. I even researched and used new vocabulary. I learned to write a one to two sentence thesis. I became a faster reader and was able to pull things out of the literature I never thought I could!

Student #2 Music helped me develop a mental process for expression. Music improved my ability to put ideas down. Music made more dynamic sentence structure development. It made it easier to write a thesis. I was better able to develop strong independent paragraphs. I think I am better able to comprehend what I read. I think my research paper writing skills have improved.

Fall 2003 English 151C-12:

Student #1 My grade has increased from other English courses in the past.   Music has helped me to relate better to literature. My writing has improved immensely. I have to say that I am now more confident writing a paper than before entering this classroom. My thesis statement ability has improved form previous courses. I now understand what a thesis is and the importance of identifying thesis throughout a paper. Paragraphing has improved and have extended in length. Reading comprehension has improved because I can now relate Shakespeare and other literature to modern music. I have used cut and paste in this course more than most of my English courses. Music in the English classroom setting is amazing. It not only helps people relate to literature, but also motivates students to participate and like the subject.

Student #2 My grade has improved. I received a B+ in my last course, but we never wrote things that were so in depth. I have definitely improved. The class to me was challenging, but in a good way. After I realized the connections, writing became easy for me. I enjoyed writing! When I researched the songs I was going to use, I found that there were awesome connections between the music and the literature we were studying. After I realized the connections, I ran to the computer and started typing. I realized my vocabulary improved after I put the paper together. I said to myself, "Wow, I wrote that sentence!"

Fall 2003 English101C-03

Student#1 Before coming into the Music College English Classroom, I was forced to write on subjects that I really didn't give a hoot about. Relating music to different subjects kept my interest and let me be free. I do more research now because relating music to other subjects compels you to do research. The drafts are well stretched out and we weren't loaded with work. I think my vocabulary is better due to music. Especially writing papers in this class, I strive to find better words to use. I think that this class was different. It makes me strive to relations with subjects to other subjects. It made me want to work hard. Think that music in the college classroom is a great idea. It's fun to work with.

II. MI Theory

An instructor for colleges in the Lehigh Valley, lecturer on art and writing in the State of Pennsylvania, and the recipient of the Northampton Community College Project Aware Outstanding Service Award 1978, I am also a reviewer for Kappa Delta Pi International Honor Society in Education Editorial Department. My article entitled "Pairing Linguistic and Music Intelligences" was published in the Kappa Delta Pi Record. Vol. 41, No. 3, Spring 2005, pages 128-130.

My dissertation, which is in the collections of seven libraries in the United States, was a case study survey research experiment that performed in 2003 to investigate the significant sustained performance of Northampton Community College students in technology enhanced classrooms from January 2003 to December 2003. I used single post test only to study relationships of dependent variables to the use of modern songs with lyrics and their pedagogical application in the college English technology classroom. Ultimately

I attempted to demonstrate a new framework for teaching college English through a combination of naturalistic observation and classification of variables. The subjects in the study were 70 students in two different sections of English I and three different sections of English II.

In 2004, statistical results of my case study research suggest that pairing music and linguistic intelligences in the college classroom improves students' grades and abilities to compose theses statements for research papers in courses that emphasize reading and writing skills (DiEdwardo 2004). Grounded in educational theories as well as offering practical applications, pairing two intelligences advances student potential. Based on Multiple Intelligence Theory or MI Theory (Gardner 1993) and participation as precedent to learning (Bateson 1975), pairing music with linguistic activities suggests that integrating music into college linguistic classrooms and virtual classrooms enables students to learn to read and write. Listening to instrumental music as well as songs with lyrics before and during lectures, presentations, and virtual lecture asynchronous lessons offers practical solutions to teachers.

In my initial study in 2003, I refined Howard Gardner's (1993) definition of music as a "separate intellectual competence" and compared music intelligence to linguistic intelligence. Furthermore, presently, in the distance learning educational platform, I suggest that through acknowledgement of MI Theory, educators infuse cognitive abilities of students by helping them "think" as one of my students suggested. Electronic reader response journaling, group activities, music as catalyst, processfolio, and writing process theory enlighten their linguistic, logical-mathematical, musical, special, bodily kinesthetic, naturalistic, personal and existential intelligences.

Essay 2

Coffeehouse: Live Your Writing Experience in College English

Students who read daily succeed in college. Your book is your life in an English class experience. Get to know the books required and use them daily as a part of life. Read the works and pick your favorite writers to use for research projects. Ask if you can write on writers who are not represented in the required texts.

Students who write daily succeed in college writing classes. Use every activity that you come across in the class experience including telephone or real face to face conferences, chat experiences, group talks, coffeehouse days, presentations of others, or web searches. ALL these activities are the stuff of writing as we write about our lives. For instance, a paper in English may be a review of a coffeehouse day and an analysis of student presentations.

Write about the lectures, web experiences like video streaming searches and reviews, group research library searches, movies you see on weekends, theatre, field trips to art museums or to your relative's home, library lessons online or in the library, group experiences, class happenings, chats, or write poetry fiction drama within the essay or author personality integration such as a letter to your author or a letter to your friend about your English study. Live your writing and you will succeed in college English class.

Write a play with your group or write a newspaper article to publish in a class newsletter or a class website. Develop a class website or a class power point series.

After all, you are the writer now and you are making history.

Chats can be coffeehouse days where you write about your experiences in writing and share thoughts on your own life as writing topics or to add to writing topics. I save blackboard chats; they are documents you can refer to as citations or you can use as preliminary parts of papers. Chats can be poetry fiction drama presentations where students post a short set of web sites or literary terms or parts of writing and gain discussion journal points or use the experience actually to write a paper. Papers can be ABOUT chats. Live internet experiences are life and we write about life or about others who understand aspects of the human condition. Blogs are live writing process as well. Create a group blog and use that for a presentation or a team paper.

First of all, conversations lead to writing process abilities. A chat can proceed through a series of conversations that can become a paper. A private journal can be a part of any writing class and can be any type of writing that helps the student move forward with literary language skills. Write about writing or your experiences in class. Use your actual feelings of self-doubt and growth to write papers. Add details by reference to the voices of the writers of your chosen library resources for your Works Cited and alluding to our text book readings. In life writing or writing about our own life stories in self discovery journaling or short stories for allusions in papers, we learn to listen carefully to events and our own responses to literary concepts. We listen. The critic is analytical, logical, sequential, organized, on track. We learn to respond to literary happenings through how we feel, not with statements of right and wrong. Learning to respond to poetry fiction drama or non-fiction allows us to give others truly helpful feedback and the creator critic in ourselves.

Life story writing can fit into our English class we are guided by quests of those who have come before us and we have accomplished meaningful tasks in life so we have a sense of purpose. Therefore, we can write about topics presented in English class through a life story perspective. Begin an essay with your own story. Filter the story as a metaphor throughout the essay or research paper. Storytelling is the highest order of language as it is the kind of tradition which makes myths important. If students write stories, select those that are vivid, and include them as journey ideas in papers, they are finding their own voices to begin their own writing path and their own "hero or heroine journey." Use dramatic techniques to write the life story and you will see you writing soar. Ultimately, writing is a personal journey determined by the student's mind, spirit, emotional ability to try to find a voice and to hear the inner voice and listen. Therefore, writers listen to their own voices first then place the answers on the page.

Essay 3 *Coffeehouse*

A coffeehouse is an informal presentation in the classroom about essay 1 or related presentations based upon literary study during the fall of 2007 in freshman English. Dr. DiEdwardo provides snacks and student may bring beverages.

Suggestions for presentations:

1. Discuss group dynamics if you are in a group that is working on strategies. For example, Lauren's group will suggest ideas for essay 2 on film based on comparison of film to book and chats as conferences.
2. Schedule a three minute visit to a different class section by DiEdwardo to present a short conversation about your essay 1 strategy.
3. show digital photos of a special place or project or aspect of your place essay
4. select passage from required text and ask me to copy page (two days before event) to discuss with class
5. Reflect on process and design your own life history project about a living or deceased person. Outline can be case history research. Interview friends or family members.

6. How do you write? Design autobiographical essay 6 pattern and begin to formulate ideas. Show a website.
7. Demonstrate one sound bite of Hispanic Literary Tradition (concentration is Native American Literary History, Tradition, Genre) through video stream or online media like news transcript, podcast, storytelling, framing life histories, or remembrances about your own home town.

Classes will celebrate art, film, music, literature, photography, or poetry through a festival party. Students bring beverages and may share optional informal presentations such as a narrative author impersonation draft or analytical essay draft. Students may write a letter to a song writer or poet, an essay about a song writer or poet, or a poem or song within an essay as well. Some students pretend to BE their poets or song writers and present a short power point or author impersonation in or not in costume. My favorite was the day a student dressed like Ben Franklin (in a rented costume!) to share his writings.

IV. Student Writing

Experience is the ink with which the story of life is written. Our cognizence and definition of reality are created by the events we encounter in everyday life. Without a daily stream of events, not only would life be incredibly boring, but in fact it would not exist. At no point in time can anyone be doing nothing, except for after death. At this point, however, life no longer exists. It is for this reason that human beings have placed such innate importance on activity. A lack of activity simulates a death-like state which stimulates people to engage in active lifestyles. Whenever a person's senses are engaged, he or she is generally having an pleasant experience. Increasing the number of senses engaged also increases the quality of the experience. It is for this reason that students generally find classes with many different mediums of information transference more enjoyable than traditional teaching methods.

The students in Professor DiEdwardo's classes were able to experience this enriched learning environment during the days she entitled "Coffeehouse Days." These were days where students were able to present their ideas for upcoming assignments in unique and individual ways. Various mediums of presentation included video, music, art, photography, advertisement and other visual aids such as PowerPoint presentations. The students were able to use any type of medium they wished for this presentation. It is not surprising that many students chose to employ various types of video media to convey their ideas. This is most likely because students recognize that videos stimulate the most senses and are thus more interesting and provide a better experience. Teachers would be wise to observe this phenomenon because it is an honest representation of what students are looking for in a class and if a teacher can adapt to this, he too will have a more enjoyable experience because there will be a positive interaction between the students and the teacher.

Although videos can be extremely effective modes of communication, it is important to consider other mediums due to their cultural significance. Perhaps one of the most important attributes of Professor DiEdwardo's "Coffeehouse Days" is the way in which they promote the consideration of a wide variety of cultural significant art forms. The synergy of technological and traditional forms of communication in the classroom on Coffeehouse days is awe-inspiring. The presentation of digital video followed closely by the demonstration of antique photographs creates a remarkable juxtaposition. This highlights both the advancement and the nostalgic tendencies of the human race. The idea of Coffeehouse Days should be adopted across the field of education because it would allow for a more complete and richer educational experience to students throughout the world. In fact there should be a new type of teaching which employs the basic ideas of Coffeehouse Days to extend them to Coffeehouse Courses or something of the like so that students may learn in a richer way all the time.

An additional benefit of the environment provided to students on Coffeehouse Days is the ability to learn from their peers. As exemplified by Peer Pressure, people in a common group have a great influence upon one another. If this impressive influence could be channeled into a way of teaching, it could be a very significant educational tool. The beginning of the phenomenon can be seen during Coffeehouse Days as students present each other with wide varieties of information and unique ideas to expand the knowledge of their classmates. This free exchange of information allows for significant and insightful discussions and revelations. This in turn provides students access to a richer and more complete education.

Coffeehouse Days are an invaluable contribution to the classroom and should be a more common occurrence. It is arguable that these days are the most important days of the entire course due to the immense educational value they hold. Their significance stems from the experience they provide. By stimulating senses, incorporating different mediums of communication, and providing for effective education, Coffeehouse Days offer the most pleasurable experience all term. This concept should be adopted universally so that learning becomes an enjoyable experience and thus cause education to come to life.

V. Mastery in Teaching through Use of Music to Transform Computer Intensive Classes

Conferencing

Email, telephone, Messaging, AOl Instant Messenger:

Student: Hi, it's … do you have a moment since we are on different time schedules I'd like to ask a question please? Hi I was just writing u an email.Thanks for getting on with me.I'm confused w/regard to Final Exam and Processfolio.Also with regard to Library Resources. All I'm using is the internet right now.

HumanSyllabus: processfolio is the collection of the work on connectweb

HumanSyllabus: make a page for the posts that states that you completed drafts and discussions for all 4 papers

HumanSyllabus: library resources are the sources in your works cited for papers

Student: For example, all the sources I've used for all my papers you want listed on a piece of paper?

HumanSyllabus: no the sources for the papers are in works cited

HumanSyllabus: the places where I will find the papers should be listed

Musicality Contextual Framework

1. Define Purpose of Lesson
2. Identify Transformational Qualities
3. List Transformational Goals
4. Music Selection or Selections
5. Literary Work or Works
6. Contextual Patterns of Music and Literature
7. Annotation of Song or Songs and Literary Work or Works
8. Metaphors in Context of Music and Literary Works
9. Compare and Contrast Music and Literature
10. Context and Music Applied to Research
11. Types of Prose

    To Narrate

    To Explain

    To Describe

    To Define

    To Compare and Contrast

    To Analyze

    To Argue
12. Electronic Reader Response Journal Discussions
13. Technology Interactions

14. Group

>   Discussions

>   Peer Critique

>   Project

15. Writing Process

>   Brainstorming

>   Clustering

>   Thesis Statement

>   Outlining

>   Drafting

16. CULTURE Student Gains Self Knowledge Evolution of Transformation

17. PRESENT Portfolio

VII. Using Modern Music Lyrics To Teach Writing

Chart of Literary Skills and Songs

	Song Name	Artist
Biblical Imagery	"My Sacrifice"	Creed
	"Signs"	
	"Stand Here With Me"	
	"Weathered"	
	"Freedom Fighter"	
Biography	"When It Hurts So Bad"	Lauryn Hill
	"Tell Him"	
	"Superstar"	

Characterization in Hamlet	"Anything But Ordinary"	Avril Lavigne
Climax	"If I Am"	Nine Days
Closure	"These Days"	Jennifer Page
Code	"Beautiful Day"	U2
Cohesiveness	"The Rising"	Bruce Springsteen
Colloquialism	"Crush"	Mandy Moore
Content	"Days of Our Lives" "Die Born"	Day of the New
Definition	"Complicated" "Nobody's Fool" "Sk8er Boi" "Losing Grip"	Avril Lavigne
Diction	"Unwell"	Matchbox Twenty
Dramatic Monologue	"Hotel Paper" "Are You Happy Now?" "Find Your Way Back"	Michelle Branch
Existentialism	"You're a God" "You Say" "We Are" "Great Divide" "Everything You Want" "Best I Ever Had"	Vertical Horizon
Free Verse	"Self Evident"	Ani DiFranco
Hamartia	"Hello Birmingham"	Ani DiFranco
Harlem Renaissance	"Where is the Love"	The Black Eyed Peas

Imagery	"A Thousand Miles"	Vanessa Carlton
Language	"Walk This Way"	Aerosmith
Metaphor	"3 A.M. " "All I Need" "Bent" "Black and White People" "Bright Lights" "Cold" "Could I Be You" "Crutch" "Disease" "Feel" "Hand Me Down"	Matchbox Twenty
Meter	"The Flower That Shattered the Stone"	John Denver
Mood	"Possibly Maybe"	Bjork
Muse	"Drops of Jupiter"	Train
Narrative	"Road Trippin'"	Red Hot Chili Peppers
Parallelism	"Absolutely (Story of a Girl)" "So Far Away"	Nine Days
Plot	"What's My Age Again" "Rock Show" "Don't Leave Me"	Blink - 182

Poetic Foot	"Slide"	Goo Goo Dolls
	"Black Balloon"	
Poetic License	"83"	John Mayer
	"City Love"	
	"Back to You"	
Point of View	"Warning"	Incubus
	"Mexico"	
	"I Miss You"	
	"Drive"	
	"11A.M."	
Prose Poetry	"You Had Time"	Ani DiFranco
	"Both Hands"	
Prosody	"Change the World"	Eric Clapton
Repetition	"Time"	Hootie and the Blowfish
Rhythm	"Collective Soul"	Skin
Sonnet	"If You're Gone"	Matchbox Twenty
	"Leave"	
	"Mad Season"	
	"Real World"	
	"Rest Stop"	
	"Soul"	
	"Stop"	
Symbolism	"The Rising"	Bruce Springsteen

Syntax	"Adrienne"	The Calling
	"For You"	
Thesis	"Everyday"	Dave Matthews Band
	"Where Are You Going"	
	"The Space Between"	
	"So Right"	
	"I Did It"	
	"Grey Street"	
	"Fool To Think"	
	"Dodo"	
Tone	"Falls on Me"	Fuel
	"Hemorrhage (In My Heart)"	
Tone	"Never Again"	Nickleback
	"Someday"	
Voice	"Bring Me To Life"	Evanescence
	"My Immortal"	
Wit	"Stairway to Heaven"	Led Zeppelin
Writing Process	"Don't Know Why"	Norah Jones
	"Seven Years"	
	"Cold Cold Heart"	
	"Feelin' The Same Way"	
	"Come Away With Me"	
	"I've Got To See You Again"	
	"One Flight Down"	

References

Bateson, Mary Catherine.(1975).Mother-Infant Exchanges: The Epigenesis of Conversational Interaction. In D. Aaronson and R.W. Rieber (Eds.), *Developmental*

Psycholinguistic and Communication Disorders. Annals of the New York Academy of Sciences, 263, 101-113.

Bateson, Mary Catherine. (1994). *Peripheral Visions, Learning Along the Way*. New York: Harper Collins.

Berman, Jeffrey. (1994). *Berman Diaries to an English Professor: Pain and Growth in the Classroom.* Massachusetts: University of Massachusetts Press.

Bogarad, C.R. and Schmidt, J.Z. (2002). *Legacies.* New York: Harcourt.

Burrows, D. (1990). *Sound, Speech, and Music.* Amherst: University of Massachusetts Press.

Campbell, Don. (2001). *The Mozart Effect.* New York: Avon Books.

Demers, M. G. (1988). *Dictionary of Literary Themes and Motifs.* New York: Greenwood Press.

Bredendieck Fischer, Gwen. (1999). *Developing Students' Adaptive Learning Skills.* College Teaching. Vol.47, Issue 3.

DiEdwardo, Maryann. "Pairing Linguistic and Music Intelligences" Kappa Delta Pi Record. Vol. 41, No. 3, Spring 2005, pages 128-130.

Gardner, Howard. (1993). *Multiple Intelligences.* New York: Perseus Books.

Gardner, Howard. (1983). *Frames of Mind.*New York:Basic Books.

Giannone, R. (1968). *Music in Willa Cather's Fiction.* Lincoln: University of Nebraska Press.

Gibson, Joanna.(2002). *Perspectives, Case Studies for Readers and Writers.* New York: Addison, Wesley and Longman.

Graziano, Anthony, Raulin, Michael. (2000). *Research Methods, A Process of Inquiry.* Boston: Allyn and Bacon.

Hardy, Miriam and Hardy, William. (1977). *Essays on Communication and Communication Disorders.* New York: Grune and Stratton.

Iudin-Nelson, Laurie June. (1997). "Songs in the L2 Syllabus Integrating the Study of Russian Language and Culture." Diss. U of Wisconsin, Madison.

Johnson, Jean. (1997). *The Bedford Guide to the Research Process.* New York: Bedford/St. Martin.

Kane, Jeffrey. (1999) *Education, Information, and Transformation, Essays on Learning and Thinking.*New Jersey: Prentice Hall.

Kramer, Daniel J. (2001) ADFL Bulletin. "A Blueprint for Teaching Foreign Languages AndCultures through Music in the Classroom and on the Web."

Krashen, Stephen D.(1989). *Language Acquisition and Language Education. 2nd ed.* Language Teaching Methodology. New York: Prentice.

Krashen, S.D.(1983).*The din in the head, input, and the language acquisition device.* Foreign Language Annals, 16, 41-44.

Neeley, Stacia Dunn.(2001). *Academic Literacy.* New York: Addison, Wesley Longman.

Palmquist, Mike. (2003). *The Bedford Researcher.* New York: Bedford/St. Martin.

Potts, Mary Jo. (1998) *Teacher's Guide – AP English Language and Composition.* New York: College Entrance Examination Board and Educational Testing Service.

Ray, Julia J. (1997). "For the Love of Children: Using the Power of Music in English as a Second Language Program." Diss.U of California, Los Angeles, 1997.

Smelcer, John. (1999) *Music as a Catalyst for Responding to Literature.* Infotrieve, Inc. ERIC. College of Education, Appalachian State University. PO Box 32038, Boone, North Carolina.

Towell, Janet. (2000) *Motivating Students through Music and Literature.* 2000. Reading Teacher; v53 n4 p284-87 Dec-Jan 1999-2000.

Weaver, J. W. (1998). *Joyce's Music and Noise: Theme and Variation in His Writings.* Gainesville, FL: University Press of Florida.

Will, Howard. (1997) *Pairing Music with Literature.* Teaching Music; v4 n5 p35-37 April.

Annotated Bibliography

Bateson, Mary Catherine.(1975).Mother-Infant Exchanges: The Epigenesis of Conversational Interaction.In D. Aaronson and R.W. Rieber (Eds.), *Developmental Psycholinguistic and Communication Disorders. Annals of the New York Academy of Sciences*, 263, 101-113.

Bateson presents her theories of coparticipation which are the basis for Dr. DiEdwardo's Dissertation. Initially, the infant mother relationship defines coparticipation and creates the environment necessary for learning to begin. Sound by both mother and child relay conversations that are the essence of coparticipation.

Campbell, Don.(2001).*The Mozart Effect*. New York: Avon Books.

In this important book, Don Campbell (2001) claims that music can help heal a large number of illnesses and problems such as acute pain, aggressive and antisocial behavior, aids, chronic fatigue syndrome, colds, depression, schizophrenia, substance abuse, tooth problems, trauma, and writer's block. He backs up his theories with research or specific examples and offers practical applications. Campbell's theory applies to the educational setting in multiple ways.

DiEdwardo, Mary Ann. (2004). "Music Transforms the College English Classroom." Diss. California Coast University, Santa Ana, California.

This document in bound format is available through California Coast University, Santa Ana, CA; Northampton Community College; Lehigh University; Pennsylvania State University, University Park, PA; Harvard University; DeSales University; and Montgomery County Community College.

Gardner, Howard. (1993). *Multiple Intelligences.* New York: Perseus Books.

The major force behind Dr. DiEdwardo's dissertation and current writing, this work engages the reader to change current education practices.Harvard University School of Education offers courses in this new important theory.

Iudin-Nelson, Laurie June. (1997). "Songs in the L2 Syllabus Integrating the Study of Russian Language and Culture." Diss. U of Wisconsin, Madison.

A fascinating study with focus on the "Din" and other intricate activities of the human brain that involve learning and the relationship between music and linguistic intelligence.

Kane, Jeffrey.(1999) *Education, Information, and Transformation, Essays on Learning and Thinking.*New Jersey: Prentice Hall.

Dr. DiEdwardo's favorite graduate text with marvelous resources for educators.Resources at the end of all chapters provide data for research.

Kramer, Daniel J. (2001) ADFL Bulletin. "A Blueprint for Teaching Foreign Languages And Cultures through Music in the Classroom and on the Web."

Essentially, one of the most important articles on the use of music to teach language.

Ray, Julia J.(1997). "For the Love of Children: Using the Power of Music in English as a Second Language Program." Diss.U of California, Los Angeles, 1997.

Detailed and systematically presented, this work shows how to design survey research on music and linguistic intelligence. Illustrations of lessons serve to bring the actual survey alive and allow the reader to participate in the actual events.

Smelcer, John.(1999) Music as a Catalyst for Responding to Literature. Infotrieve, Inc. ERIC. College of Education, Appalachian State University. PO Box 32038, Boone, North Carolina.

Smelcer writes a short factual explanation of how he uses music to teach literature. His presentation relies on music without lyrics. Dr. Diedwardo moved the research forward to include music with lyrics to transform the literature classroom.

About the Author

I have been an adjunct faculty member for over thirty years in Lehigh Valley Colleges. In fact, I started to teach to NCC in 1975 as a tutor for the Project Aware and our NCC ESL Department. Subsequently, I was hired as a Professional Assistant in our Reading-Writing Lab where I often designed classes in research and writing. In 1978, I was the recipient of the Project Aware Outstanding Service Award. Since 1993, I have had experience working with traditional and distance learning students in need of assistance with their questions and concerns; my strong computer skills (ANGEL at Penn State, WebCT at LCCC, and Connectweb and Blackboard at NCC) where I have performed as adjunct faculty show experience with digital learning management systems and ability to manage projects and to handle multiple tasks. Currently, I provide lectures for conferences in the Northeast on topics in education such as my new theory called the Fourth "R" or remembrance in testimonials and memoir. I been lecturing in the field of curriculum design, technology based writing programs, and linguistics.

I use current technology to teach and to serve academic communities. In 2007 I performed a case study on my Penn State English I class to see how my podcasting motivated their learning styles. I also teach that catalysts help learning. I take my students to art exhibits and recently performed an art show for Penn State University at the Lehigh Valley campus called ART FOR A CAUSE.

I hold a Master's Degree in English from Lehigh University and a Doctorate of Education from an accredited Distance Learning University entitled California Coast University with successful teaching and program management experience related to writing instruction and distance learning with previous supervisory responsibilities in post-secondary institution as LCCC (Lehigh Carbon Community College) distance learning course designer for English 106-M1, the first online distance mid-semester English II class.

Article Excerpt

A fourth "R" called remembrance has changed my educational focus from what exists on the page to what can exist there with imagination and fortitude. Teach courage based upon oral histories! The concept presented discovery as I continue to teach writers of all ages in demonstrations, lectures, community groups at libraries and schools as well as classrooms in local colleges and universities in Pennsylvania. Rooted in educational theory based upon Hispanic literary traditions, historical memoir writing creates students voices. With a combination of historical facts formulated through research or real life experiences and based upon fiction writing techniques, students of all ages intertwine personal oral history into written form to center their minds to write. Students of writing from K-adult learn to write based upon inherent knowledge. When students learn to use their own voices, they interpret assignments with fresh views and confident attitudes.

Historical remembrance as a fourth "R" instills writing ability in students in pre-school classes with journaling, a fun activity. Next, the elementary student functions well within the journaling assignments with notebooks filled with imagined or real tales. These can be credited toward writing core curriculum development. As students progress through the intermediate and secondary years, they need to recall past events of their own lives but base recollections on history as part of integrated curriculum. High school students write in seminar type settings where they use the historical memoir to heal or to reflect on life as their own stories progress through events past, present and future. Gradually, the writing students who select English I and II, the first college writing experiences, may concentrate on oral history exercises followed by the oral history writing process and ultimately an oral history framework that changes the reluctant writer into the confident writer.

My first experience with history as paradigm for teaching English developed from ten years of teaching in a global community via the online tutorials. The interplay of history and English helped students write stories, oral histories, folk lore, or biblically based tales as they ventured into writing as a hobby. I held classes in my home kitchen for local young writers, ages 5-19. Soon, the writers could take a pet or a story of a grandparent and write a tale of remembrance that might be based upon history, fiction, fact, and memoir.

As a memoirist, poet, playwright, fiction writer and prose author, my discovery that writing about historical topics helps students focus, learn to create, and eventually seek writing as a hobby and college skill of choice is based upon thirty years inside the classroom. I tell my students of all ages that good writing forms a package: beginning, middle and end. In all classrooms, no matter what the age group, I teach students to form ideas through structure, action, story, solution, theme, tone and language. Students of multicultural traditions love to hear their own languages, so I form works through coding or allowing the students to use small portions of native language within the writing assignment.

Other Books by the Author

DiEdwardo, Maryann Pasda and Pasda, Patricia. *Horses about Hope, the art of myth, legend and graphic writing.* Bloomington, Indiana: AuthorHouse Publishing Company, U.S.A. 2007.

DiEdwardo, Maryann Pasda. *Music Transforms the College English Classroom.* Bloomington, Indiana: AuthorHouse Publishing Company, U.S.A. 2007.

DiEdwardo, Maryann Pasda and Pasda Patricia. *On Healing: Inspiration through Writing.* Bloomington, Indiana: AuthorHouse Publishing Company, U.S.A. 2007.

DiEdwardo, Maryann Pasda and Pasda, Patricia. *Pennsylvania Voices I, II, III, IV, V.* Bloomington, Indiana: AuthorHouse Publishing Company, U.S.A. 2007.

DiEdwardo, Maryann Pasda and Pasda, Patricia. *The Art of Trees.* Bloomington, Indiana: AuthorHouse Publishing Company, U.S.A. 2007.

DiEdwardo, Maryann Pasda. *The Fourth "R" A Book to Promote the Journey through Hispanic American Literary History to Develop Language Skills.* Bloomington, Indiana: AuthorHouse Publishing Company, U.S.A. 2008.

"Dr. DiEdwardo's book is a 'must have' for all educators, especially for those who teach students of other languages. The book is simple, and its components are easy to follow. What I find particularly compelling about this book is the idea to use authors from the students' home country to enhance self-esteem and pride, in addition to creating individual voices. Building background knowledge, and giving students a voice encourages the pupils to express themselves, and it makes teaching writing more enjoyable." Toni Velleca, ESOL Teacher

DiEdwardo, Maryann Pasda and Pasda, Patricia. *The Horse Keeper, The Healing Gifts of Painting and Writing about Horses._West Conshohocken, Pennsylvania: Infinity, 2008.

DiEdwardo, Maryann Pasda. *The Legacy of Katharine Hepburn: Fine Art as a Way of Life._Bloomington, Indiana: AuthorHouse Publishing Company, U.S.A. 2006.

DiEdwardo, Maryann Pasda and Pasda, Patricia. *Pennsylvania Voices On Healing: Cancer Survival Through Prayer Painting and Writing.* Orlando: Xulon Press, 2007.